2026 EDITION

SEO AI GROWTH SYSTEM

The Wiser Optimizer System

A Complete Digital Marketing Operating Manual

VISIBILITY	DEMAND	TRUST	CONVERSION	MEASUREMENT

"A fool with a plan can outsmart a genius with no plan any day." — T. Boone Pickens

By John Vincent Kriney

optfirst.com · firstfirms.com

Table of Contents

FOREWORD

About This Book's Origins

The first book in this series, The 8 Step SEO Plan, was published in 2014. At the time, ranking a website on Google was largely a technical exercise — structured around keywords, backlinks, and metadata. The businesses that followed the eight-step process consistently saw results. The system worked because search itself was relatively simple.

The second book, The Search Engine Marketing Manual, expanded on that foundation. It introduced paid search alongside organic and showed how aligning both channels amplifies overall performance. Readers learned that SEO and SEM are not competing strategies. They are complementary forces. When both are running cleanly, the combined output exceeds what either produces alone.

The third book, The Online Marketing Manual, broadened the lens further — exploring how search and social work together to generate a larger, more compounding impact. That book introduced the idea of marketing as a multi-channel system rather than a single-channel tactic.

Each book reflected the reality of its moment. And now, the reality has changed again — more dramatically than at any point in the last decade.

The Evolution of This System

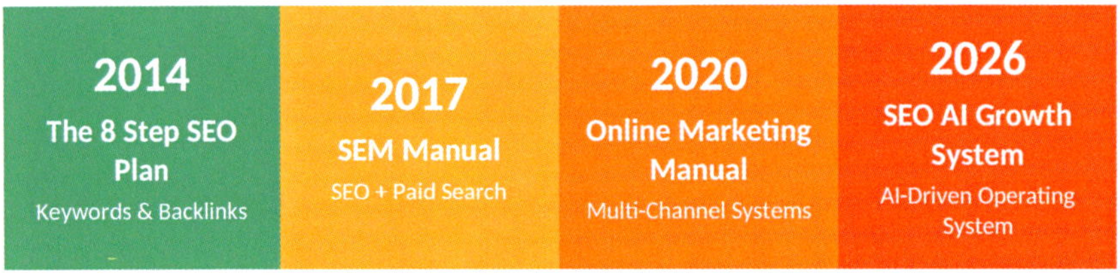

A Note Before You Begin

This is not a theoretical guide. Every framework, system, and recommendation in these pages comes directly from campaigns built, managed, and optimized across real clients in real markets — from local service providers to national ecommerce brands, from healthcare providers to law firms, from start-ups to organizations generating over $60 million in annual revenue.

The strategies here have been tested under competitive conditions. They have been refined through failure and validated through results. They are structured to be applied — not just understood.

> *"The businesses that have grown most successfully using this system have one thing in common. They did not treat it as a menu from which to choose selectively. They built the full system, connected the components, and executed with discipline. The results compounded over time."*
>
> — John Vincent Kriney

INTRODUCTION

A New Reality

There was a time when the marketing landscape was manageable because it was compartmentalized. SEO teams focused on rankings. Paid teams focused on clicks. Social teams focused on engagement. Reputation was treated as an afterthought — something to address when a negative review surfaced, not something to strategically build and protect.

Each channel had its own metrics, its own reports, its own strategy, and its own team. That model had its limitations, but it also had a certain simplicity. You could point to a single campaign, a single metric, and declare it working or not working.

That model no longer reflects reality.

Consider what happens today when someone is looking for a legal firm. They open Google and type "personal injury attorney near me." Within a fraction of a second, Google evaluates which businesses have the strongest local authority signals, the most consistent entity data, the most recent and positively-reviewed Google Business Profiles, the fastest and most mobile-responsive websites, and the content that most clearly addresses that exact query. It then generates an AI Overview — a synthesized answer pulled from sources it considers most trustworthy — before the user has clicked a single result.

If your business is not part of that answer, you may not be considered at all.

Then the user scrolls down. They see the Map Pack — a set of three local results with star ratings prominently displayed. Studies consistently show that most local clicks go to these three positions. If you are not in the Map Pack, you are invisible to the majority of local searchers.

The user clicks through to a result, reads reviews on Google, checks your website, scrolls through your social presence, and within ninety seconds, makes a judgment about whether to contact you. Every one of those touchpoints contributed to that decision.

The Modern Buyer Journey

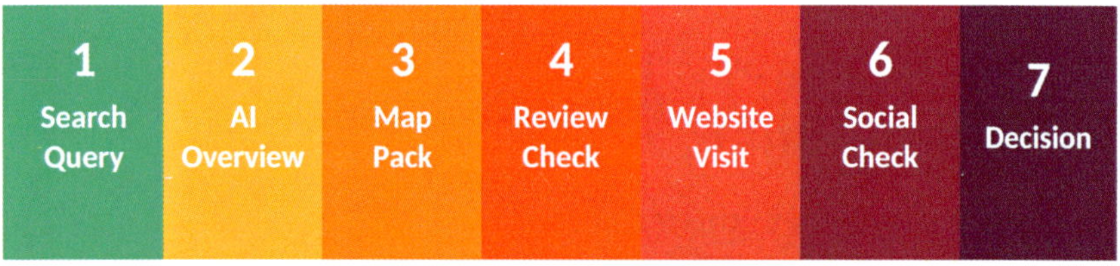

The modern buyer journey: from search query to decision in under 90 seconds.

Every Channel Influences Every Other Channel

Modern marketing operates as an interconnected system. Search engines now rely on artificial intelligence to evaluate content, intent, and trust. Advertising platforms use machine learning to determine bids, audiences, and placements in real time. Paid Social platforms reward creative that captures attention instantly. Reputation signals directly influence conversion rates. Landing page performance affects both visibility and cost efficiency. First-party data fuels optimization across every major platform.

When these systems are aligned, performance compounds. When they are disconnected, performance deteriorates.

"My life has been a product of compound interest. Nothing more. Nothing less. And nothing brilliant."

— Warren Buffett

The Wiser Optimizer Approach

At OptFirst and FirstFirms, marketing is not approached as a series of disconnected services. What gets built for clients are SEO AI Growth systems — structured around two core disciplines:

1. Building and Running Connected Campaigns

Performance-driven strategies are developed and managed across the channels that define modern growth:

- Search Engine Optimization (SEO)
- Generative Engine Optimization (GEO)
- Search Engine Marketing (SEM) — Google Ads and Performance Max
- Paid Social Campaigns
- Reputation Management
- Retargeting and full-funnel acquisition

2. Infrastructure Before Performance

Sustainable growth requires more than campaigns. It requires infrastructure. Building a high-performing marketing platform requires the following assets to be developed, integrated, and optimized:

- Websites and landing pages
- Ecommerce experiences
- Web applications and internal tools
- Tracking systems and analytics environments
- CRM integrations
- Server-side data and conversion architecture

Without the right foundation, campaigns become inefficient, difficult to scale, and unreliable to measure. With the right foundation, achieving high performance becomes predictable.

Data Across Markets Sharpens Expertise

Over almost two decades, the businesses engaged in these campaigns have spanned a wide range of industries: local service providers, national brands, ecommerce companies that scaled well over $60 million in year-one sales, major universities, and organizations operating in highly visible or reputation-sensitive environments.

What has become clear is that working across industries provides an unexpected yet powerful advantage. Patterns become clear. Wins in one vertical are transferable to clients in entirely different verticals. While every business is different, the underlying mechanics of growth are remarkably consistent.

The Objective

The objective of modern marketing is not traffic. It is not impressions. It is not clicks. The objective is measurable growth.

At OptFirst, the focus is simple: to turn visibility into action, and action into revenue. Every system outlined in this book is designed with that outcome in mind — whether the goal is to increase calls, generate qualified leads, drive ecommerce sales, or improve long-term customer value.

SYSTEM OVERVIEW

The Five-Layer Wiser Optimizer System

"Losers have goals. Winners have systems."

— Scott Adams

The system operates as five connected layers. Each layer strengthens the next. When aligned, performance compounds.

1	**VISIBILITY** Building organic and AI-driven search presence

2	**DEMAND** Capturing and creating intent through SEM and Paid Social

3 **TRUST**
Building reputation and credibility before conversion

4 **CONVERSION**
Designing systems that turn traffic into measurable revenue

5 **MEASUREMENT**
Unifying data to drive decisions and accelerate growth

The Compounding Effect

Each layer of the system feeds the next. SEO visibility increases click volume. That click volume generates conversion data. That data powers better paid advertising. Better advertising increases brand visibility. Increased visibility generates more reviews. More reviews build the trust that converts cold traffic. The system does not just produce results — it produces results that build on themselves.

Figure 1: The 5-Step Growth Process

Visibility → Demand → Trust → Conversion → Measurement

PART 1

SYSTEM FOUNDATION

Building the Infrastructure for Growth

Before campaigns, before creative, before content — you must build the foundation that supports measurement, optimization, and scalable growth. Part 1 establishes the two pillars: Conversion Architecture and Unified Analytics.

SECTION 1.1

Conversion Architecture

Designing Experiences That Turn Visibility Into Revenue

Conversion architecture is not design. It is a structured system engineered to influence behavior. Traffic does not create revenue. Conversions do. Every system in this book — SEO, GEO, SEM, Paid Social, Reputation — ultimately serves one purpose: feeding a conversion system that produces measurable outcomes.

Most businesses treat conversion as an outcome. High-performing systems treat conversion as a design variable.

The Conversion Reality

Every user that reaches your site is asking one question: Is this the right decision for me, right now? Your job is not to convince. Your job is to:

- Reduce uncertainty
- Increase clarity
- Provide evidence
- Make action obvious

If you fail to do this quickly, the user leaves. Not because they are uninterested. Because they are unconvinced. The average user decides whether to stay on a page within 8–10 seconds of arrival. That window is determined by the quality of your first impression — page load speed, headline clarity, and whether the experience matches what they expected.

The Four Variables That Control Conversion

1 Intent Alignment	2 Clarity	3 Trust	4 Friction

Experience matches what the user expected when they clicked	User understands the offer, audience, value, and next step instantly	Social proof, reviews, certifications, and guarantees reduce risk	Every additional step reduces conversion probability

The Six Layers of High-Performance Conversion Architecture

1. Message-to-Market Match

Every experience must align with the user's intent, traffic source, and stage of awareness. Your messaging must match the keywords that brought them (SEO/SEM), the ad creative they clicked (Paid Social), and their audience expectations. If messaging breaks, the system breaks.

2. Above-the-Fold Control

The first screen determines whether the user continues or exits. Within seconds, the user must see a clear headline, a defined outcome, a compelling reason, and a visible next step. This is not a branding space. It is a decision space.

3. Narrative Flow

High-converting pages follow a predictable psychological sequence: Problem recognition → Solution introduction → Proof and validation → Risk reduction → Action guidance. If the flow is broken, users disengage.

4. Trust Infrastructure

Trust must be layered, not assumed. Reviews with recency and detail, testimonials that are specific and believable, case studies that are results-based, certifications, media mentions, and guarantees. A page with fifteen trust signals converts better than a page with five.

5. Friction Elimination System

Friction is not always visible. It often exists in too many choices, poor layout hierarchy, unclear instructions, or unnecessary steps. High-performing systems reduce decisions, simplify paths, and eliminate distractions.

6. Conversion Path Engineering

Every page must answer: What is the next step, and how fast can it happen? Conversion paths should be direct, obvious, and minimal. Click-to-call for high-intent local pages, short forms (3–4 fields) for lead generation, streamlined one-page checkout for ecommerce.

First-Party Data as a Conversion Multiplier

Every conversion point should also function as a data asset. You should capture emails, phone numbers, behavioral signals, and conversion actions. This data fuels retargeting, AI optimization, and lifetime value expansion. Without data, your system cannot improve. With it, every campaign you run becomes smarter over time.

Continuous Optimization: Where Growth Compounds

Conversion architecture is never complete. It evolves through A/B testing, behavioral tracking, funnel analysis, and heatmaps. Focus testing on headlines, CTAs, layout structure, trust elements, and form friction. Small improvements compound over time. A business that improves its conversion rate from 2% to 3% has just increased revenue from the same traffic by 50% — without spending an additional dollar on advertising.

> **The Conversion Principle**
>
> When you match intent, communicate clearly, build trust, and remove friction — users convert.
>
> When users convert, data improves, platforms optimize, costs decrease, and revenue increases.
>
> Conversion is not an outcome. It is a system.

SECTION 1.2

Unified Analytics

Turning Data Into Clarity, Decisions, and Scalable Growth

If conversion architecture determines performance, analytics determines direction. Without analytics, you operate blindly, misinterpret success, and scale inefficiency. With analytics, you see patterns, identify leverage, and scale intelligently.

What Analytics Really Is

Analytics is not reporting. It is not dashboards. It is not charts. Analytics is a decision-making system. It exists to answer four questions:

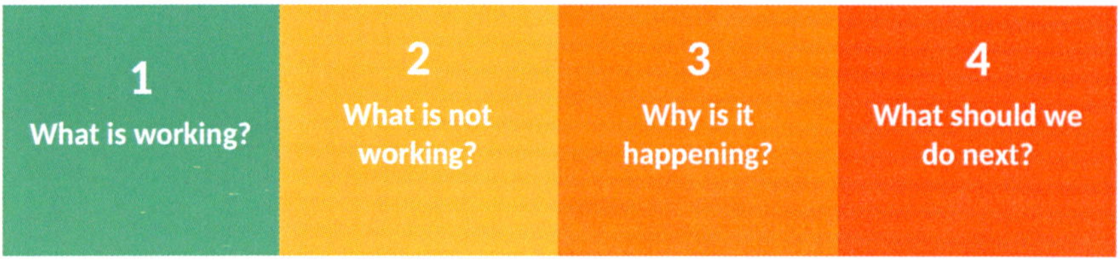

Anything that does not drive decisions is noise.

The Core Problem: Fragmentation

Most businesses operate with disconnected systems: Google Ads data, Meta Ads data, SEO tools, CRM platforms, call tracking, and analytics tools. Each system shows a partial truth. No system shows the full picture. This creates conflicting conclusions, misaligned teams, and inefficient scaling.

> ### The Cost of Fragmentation
>
> Real Example: A business sees 200 leads/month in Google Ads. The CRM shows 80 deals entered. Finance reports 30 closed clients. Which number is marketing responsible for? Without a unified system connecting all three, nobody can answer that question — and smart investment decisions are impossible.

Why Attribution Determines Everything

Users do not convert in a straight line. A typical conversion journey in 2026 might look like this:

- User searches "personal injury attorney"
- Sees organic result, clicks, reads content, leaves
- Sees retargeting ad on Facebook that evening
- Reads three reviews on Google
- Returns two days later via direct search for the firm's name
- Calls the office

If your system only tracks the final click, you attribute the conversion to branded search and conclude that everything else was irrelevant. That conclusion is wrong, and it will cause you to underinvest in the touchpoints that actually built the relationship.

Old Metrics vs. Modern Metrics

Old Metric	Why It's Insufficient	Modern Replacement
Clicks	Ignores quality and revenue impact	Customer Acquisition Cost (CAC)
Impressions	No connection to business outcomes	Return on Ad Spend (ROAS)
Traffic	Volume without conversion context	Lifetime Value (LTV)
Rankings	Visibility does not equal revenue	Revenue contribution by channel
Engagement rate	Activity does not equal performance	Conversion efficiency

Real-Time Decision Environments

Modern systems require immediate feedback. Not monthly reports or delayed analysis, but live dashboards, real-time tracking, and continuous monitoring. In a monthly reporting model, a campaign that starts underperforming on Day 3 might not be identified until Day 30. In a real-time environment, that issue is visible and addressable within hours.

The Analytics Feedback Loop

1 Data Collected	2 Insights Identified	3 Decisions Made	4 Systems Optimized	5 Performanc e Improves	6 Better Data Generated

The loop compounds — each iteration more accurate, more valuable, more actionable.

The Unified Analytics Principle

What you measure together is what you grow together. When your data is unified, patterns become visible, waste becomes obvious, and opportunities become scalable. Marketing and sales alignment is not a culture issue. It is a data infrastructure issue.

SECTION 1.3

Website & Landing Page Infrastructure

The Technical Foundation That Every Campaign Depends On

Every paid click, every organic visit, every social impression eventually lands somewhere. That somewhere is your website or landing page. Before any campaign performs, before any budget is allocated, before any content is published — your digital infrastructure must be built to receive traffic and convert it.

Most businesses treat websites as a branding asset. High-growth businesses treat them as a revenue system. The distinction determines whether marketing investment compounds or evaporates.

The Infrastructure Hierarchy

1	**Technical Foundation**	Site speed, crawlability, security (HTTPS), mobile responsiveness, Core Web Vitals compliance
2	**Conversion Architecture**	Intent-matched pages, clear CTAs, trust signals, friction elimination
3	**Analytics Layer**	GA4, GTM, conversion tracking, heatmaps, session recordings
4	**CRM Integration**	Lead capture connected to sales pipeline, automated follow-up, attribution
5	**Testing System**	A/B testing capability, multivariate testing, behavioral data loops

Page Speed: The Silent Revenue Leak

Page speed is the most measurable and most ignored conversion lever in digital marketing. Every second of delay compounds into significant revenue loss at scale.

Metric	Impact	Target
LCP (Largest Contentful Paint)	Directly correlates with bounce rate and conversion	< 2.5 seconds
INP (Interaction to Next Paint)	Affects perceived responsiveness and user trust	< 200ms
CLS (Cumulative Layout Shift)	Visual instability damages credibility and clicks	< 0.1
TTFB (Time to First Byte)	Server response; affects all subsequent metrics	< 800ms
Total Blocking Time (TBT)	JavaScript blocking main thread delays interactivity	< 200ms

Mobile-First Technical Requirements

Google indexes and ranks your mobile site. A website that passes desktop benchmarks but fails on mobile is, from Google's perspective, a failing website.

Mobile Requirements (Technical):

- Responsive layout (no horizontal scroll)
- Touch targets minimum 44x44px
- Fonts minimum 16px for body text
- No intrusive interstitials on mobile
- Images compressed for mobile bandwidth

Mobile UX Requirements:

- Lazy loading for below-fold media
- AMP or lightweight page architecture
- Click-to-call on all phone numbers
- Forms optimized for mobile keyboards
- Viewport meta tag correctly configured

Conversion Rate Optimization (CRO) Framework

CRO is not guessing. It is a structured process of forming hypotheses based on data, testing those hypotheses, and scaling what works.

1. Analyze

Use GA4, heatmaps, and session recordings to identify where users drop off.

2. Hypothesize

Form specific, testable hypotheses with expected directional outcomes.

3. Prioritize

Score tests by potential impact × confidence ÷ effort.

4. Test

Run A/B tests with 95% statistical significance threshold.

5. Implement

Roll out winners; document losers — both inform future tests.

6. Scale

Apply winning principles across similar pages.

> **The CRO Principle**
>
> The fastest path to revenue growth is rarely more traffic. A site converting at 4% generates twice the revenue of a site converting at 2% on identical traffic. CRO compounds every other investment you make in marketing.

PART 2

SEO & GEO

Search Engine & AI Optimization

Modern search is powered by AI. Part 2 covers the three integrated layers — On-Page, Off-Page, and Generative Engine Optimization — and demonstrates their impact through three real-world case studies.

PART 2 OVERVIEW

From SEO to GEO: The Evolution of Search

Search has fundamentally changed. It is no longer just a list of blue links, a keyword-matching engine, or a ranking-based system. It is now an AI-driven interpretation engine, a summarization system, and a decision-making assistant. This introduces a new layer of practice: Generative Engine Optimization (GEO).

SEO determines whether you rank. GEO determines whether you are chosen.

The Three Integrated Layers of Modern SEO

1	2	3
ON-PAGE SEO	OFF-PAGE SEO	GEO
Foundation: how your website communicates with search engines, AI systems, and users.	Validation: how the rest of the internet references, mentions, and trusts you.	AI Extraction: how AI systems interpret, summarize, and present your brand.

These three layers must align. If they do not, you may rank but not convert, be visible but not trusted, or be cited incorrectly by AI systems.

SECTION 2.1

On-Page SEO

Building Machine and Human Clarity

On-page SEO is not about keywords. It is about clarity — for search engines (crawlability, structure), AI systems (entity understanding, summarization), and users (experience, intent satisfaction). When all three align, performance compounds.

Core Web Vitals: Performance That Feels Fast

1	2	3
LCP Largest Contentful Paint Main content visible. Target: < 2.5 sec	**INP Interaction to Next Paint** Page responds to input. Target: < 200ms	**CLS Cumulative Layout Shift** Visual stability. Target: < 0.1

The 10-Step On-Page SEO Framework

#	Name	Focus	Key Action
1	Index & Crawl Health	Accessibility	Ensure site can be accessed, crawled, and indexed
2	Intent Mapping	Purpose	Define a clear goal for every page
3	Content Structure	Clarity	Logical hierarchy and semantic organization
4	Performance Optimization	Speed	Improve LCP, INP, CLS — all Core Web Vitals
5	Structured Data	Machine clarity	Implement schema markup across key page types
6	Conversion Architecture	Revenue	Guide users toward meaningful actions
7	Technical Tag Integrity	Tracking	Ensure accurate tags, metadata, and canonicals
8	First-Party Data Strategy	Intelligence	Capture and use behavioral data ethically
9	Engagement Optimization	Behavior	Improve layout based on real user data
10	GEO Preparation	AI Readiness	Structure content for AI extraction and citation

E-E-A-T: Trust as a Ranking System

E-E-A-T stands for: Experience, Expertise, Authoritativeness, Trustworthiness. It is not a single ranking factor — it is a framework Google uses to evaluate overall quality. Trust is not declared. It is demonstrated.

1	2	3	4
Experience	**Expertise**	**Authoritativeness**	**Trustworthiness**
First-person accounts, case studies, original data	Named authors, credentials, topic authority	External mentions, links, industry recognition	Transparency, accurate info, security signals

Structured Data: Speaking the Language of Machines

Structured data is how you tell search engines what your content means, help AI systems interpret relationships, and enable rich results.

Common schema types: Organization, LocalBusiness, Product, Review, FAQ, Article, LegalService, MedicalBusiness. Without structured data, your content is interpreted. With it, your content is understood.

SECTION 2.2

Off-Page SEO

Building Authority and Trust Across the Web

Off-page SEO answers one question: Does the internet trust you? It is how search engines validate your credibility, your reputation, and your relevance. Without off-page signals, your site is unverified.

The Shift: From Links to Entities

Old Approach: Link Building	Modern Approach: Entity Building
Acquire backlinks	Build brand recognition across the web
Focus on domain authority score	Build consistent entity data everywhere
Guest posts for link juice	Digital PR for editorial mentions
Directory submissions	Strategic platform presence in your category
Keyword anchor text manipulation	Topic associations and brand signals

Figure 2: The 8-Step Off-Page SEO Plan

Migrate · Validate · Originate · Replicate · Differentiate · Calculate · Investigate · Duplicate

The 8-Step Off-Page SEO Framework (2026)

1	MIGRATE	**Unify Data, Access, and Ownership** All systems must operate from a single source of truth: analytics platforms, Search Console, CRM, social profiles, and ad accounts.

2	VALIDATE	**Establish Real-World Credibility** Consistency across business name, address, phone number, domain, and brand identity. Conflicting data weakens trust.

3	ORIGINATE	**Build Your Entity Footprint** Establish presence across core platforms (Google, LinkedIn, Facebook), industry-specific directories, review platforms, and aggregators.
4	REPLICATE	**Map and Match Authority Signals** Analyze competitor backlinks, mentions, and citations. Close authority gaps. Avoid spam tactics.
5	DIFFERENTI ATE	**Create Signals Worth Referencing** Original content, research, newsworthy assets, thought leadership. Modern authority is earned when others cite you.
6	CALCULATE	**Measure Authority as a System** Track brand mentions, review velocity, sentiment trends, branded search volume, and entity recognition.
7	INVESTIGAT E	**Analyze Performance and Adjust** Continuously evaluate ranking movement, competitor changes, content performance, and reputation signals.
8	DUPLICATE	**Scale What Works** When something works, expand it, repurpose it, and reapply across platforms. Authority compounds through consistency.

Reviews: The Most Underrated SEO Signal

Reviews influence local rankings, click-through rates (star ratings in SERPs), conversion rates (social proof), and AI summaries (sentiment is extracted into AI overviews). Consistency matters more than spikes.

Review Best Practices

- Recency: A review from last week is worth significantly more than one from last year.
- Volume: Aim for 4–8 new reviews per month minimum.
- Detail: Specific narratives outperform "Great service!" by an order of magnitude.

- Consistency: Steady weekly cadence beats occasional spikes.
- Response Rate: Always respond within 24–48 hours.

SECTION 2.3

Generative Engine Optimization (GEO)

Optimizing for the AI Era

GEO is the practice of optimizing your content and brand so that AI-powered systems — including Google's AI Overviews, ChatGPT, Perplexity, and voice assistants — accurately understand, extract, and reference your content.

Before a user clicks any search result, AI systems now present a synthesized answer drawn from sources they consider authoritative. If your content is not structured to be extracted by these systems, you are invisible before the click even happens.

How AI Systems Evaluate Content

1	2	3	4	5
Clarity	**Structure**	**Authority**	**Consistency**	**Facts**
Simple, direct language easily understood	Organized headings, sections, summaries	Citations, credentials, external validation	Aligned messaging across all platforms	Specific claims with supporting evidence

The 6 GEO Optimization Principles

1. Use Clear Definitions

Explain concepts simply and directly. AI systems extract definitions — if you define your service clearly, your definition is more likely to be cited in AI summaries.

2. Structure Content Logically

Use headings, sections, and summaries. AI systems are designed to parse structured content. Clear H1 + H2 sections + closing summary is significantly more extractable than flowing prose.

3. Build Entity Relationships

Connect topics, brands, and concepts. When your content references your location, specialty, credentials, and associated concepts, AI systems build a richer understanding.

4. Provide Concise Answers

Answer questions directly within the first paragraph. The lead answer is what gets extracted into AI summaries.

5. Support Claims with Evidence

Use facts, examples, and data. Unsupported claims are not cited. Evidence-backed claims are extracted.

6. Maintain Cross-Platform Consistency

Ensure messaging is aligned across your website, Google Business Profile, social profiles, and directories. Inconsistency results in incorrect AI descriptions.

> **GEO and Brand Control**
>
> If you do not define your narrative, AI systems will — based on whatever they find about you across the web. A business with clear, structured, authoritative content controls how it is described in AI summaries. A business with thin or ambiguous content is described by whatever other sources say about it — including its competitors.

SECTION 2.4

Local SEO: The Map Pack System

Dominating Local Search for Service-Area Businesses

For any business that serves customers in a geographic area — urgent care clinics, law firms, home service providers, restaurants, retail stores — local search is the highest-intent, highest-conversion traffic source available.

A user who types "emergency plumber near me" at 11pm is not researching. They are ready to buy. The business in the Map Pack wins that customer.

How Google's Local Algorithm Works

1 RELEVANCE	2 DISTANCE	3 PROMINENCE
How well your business matches the search query — determined by GBP categories and content alignment.	Physical proximity to the searcher. Cannot be directly controlled but can be addressed through location pages.	How well-known and trusted your business is — determined by reviews, citations, backlinks, and mentions.

Google Business Profile: The Local SEO Centerpiece

Your GBP is the most important single asset in local SEO. It controls what appears in the Map Pack, the Knowledge Panel, and Google Maps. An optimized GBP can drive more leads than your website.

GBP Optimization Checklist

- Business Name — Use your exact legal business name. No keyword stuffing.
- Primary Category — The most important field. Choose the most specific, most relevant primary category available.
- Secondary Categories — Add all relevant secondary categories.
- Business Description — 750 characters of keyword-rich, compelling copy.
- Photos & Videos — Minimum 10 photos. Businesses with photos receive 42% more direction requests.
- Services & Products — List every service with descriptions and pricing.
- Q&A Section — Pre-populate with your own questions and answers.
- Posts — Publish weekly GBP posts.
- Review Management — Respond within 24–48 hours.

Citation Building: NAP Consistency

A citation is any mention of your business Name, Address, and Phone number across the web. Citation inconsistency — even minor differences like "St." vs "Street" — dilutes trust signals and suppresses local rankings.

Citation Tier	Examples	Priority
Tier 1 — Core Platforms	Google Business Profile, Apple Maps, Bing Places, Facebook	Immediate — Week 1
Tier 2 — Data Aggregators	Localeze, Acxiom, InfoUSA, Neustar	Week 1–2
Tier 3 — General	Yelp, Yellow Pages, Foursquare, BBB, Angi	Week 2–4
Tier 4 — Industry-Specific	Healthgrades, Avvo, Houzz	Month 2+
Tier 5 — Local	Chamber of Commerce, neighborhood directories	Ongoing

SECTION 2.5

Content Marketing & Thought Leadership

Building Authority Through Consistent, Valuable Content

Content marketing is the long-game channel. It does not generate leads overnight. But done consistently over 12–24 months, it creates a self-reinforcing authority engine: more content means more rankings, more rankings mean more traffic, more traffic means more conversions, more conversions mean more data, and better data fuels better content.

In the GEO era, content serves a dual purpose: ranking in search and being cited in AI-generated summaries. The same principles that make content rank well — clarity, structure, authority, specificity — are the exact principles that make content extractable by AI.

The Content Hierarchy

1	**Pillar Pages**	2,000–5,000 words. Comprehensive guides covering a topic in full depth. Rank for broad keywords, serve as content hubs.

2	**Cluster Pages**	800–1,500 words. Specific subtopic pages that link to the pillar page. Rank for long-tail keywords and feed the pillar.
3	**FAQ Content**	300–600 words. Short, direct answers to specific questions. Optimized for featured snippets, voice search, AI extraction.
4	**Thought Leadership**	1,000–2,000 words. Opinion pieces and original perspectives. Build E-E-A-T signals and drive authority.

The Content Calendar Framework

Consistency beats intensity. A business that publishes one well-researched piece per week outperforms a business that publishes 20 pieces in January and nothing for the next six months.

Content Type	Frequency	Primary Goal
Pillar Page / Guide	1 per quarter	Rank for broad keywords; build topical authority
Cluster Article	2–4 per month	Rank for specific queries; feed pillar pages
FAQ / Quick Answer	4–8 per month	Featured snippets, voice, AI overviews
Case Study	1 per month	Social proof + E-E-A-T + sales enablement
Thought Leadership	1–2 per month	Brand authority, media citations, networking
Video Content	2–4 per month	Engagement, social reach, YouTube SEO

The Content Marketing Principle

The single most important content principle: answer the question your customer is already searching for. Not the question you wish they were asking. The question they type into Google at 11pm

Content Type	Frequency	Primary Goal
when they have a problem.		

SECTION 2.6

Technical SEO Deep Dive

The Foundation That Makes All Other SEO Possible

Content quality, authority building, and GEO optimization all depend on one prerequisite: Google must be able to access, crawl, and understand your website. Technical SEO is the engineering layer that ensures no barriers exist between Google's crawlers and your content.

Technical issues are silent killers. A site with an incorrect robots.txt file can be entirely de-indexed without the business owner knowing. These issues do not announce themselves — they require systematic auditing to discover.

The Technical SEO Audit Framework

Category	What to Check	Common Issues	Priority
Crawlability	robots.txt, XML sitemap, crawl budget	Pages blocked; sitemap not submitted	Critical
Indexation	Search Console coverage report	Noindex on important pages; errors	Critical
Site Architecture	URL structure, internal linking, depth	Pages 4+ clicks from homepage; orphans	High
Page Speed	Core Web Vitals (LCP, INP, CLS)	Render-blocking resources; images	High
Mobile Usability	Mobile-first indexing compliance	Different content; small tap targets	High
HTTPS /	SSL certificate,	Expired SSL; HTTP internal	Critical

Category	What to Check	Common Issues	Priority
Security	mixed content	links	
Structured Data	Schema markup validity	JSON-LD errors; missing properties	High
Canonicalization	Canonical tags, duplicate content	Self-referencing; www mismatches	High
Redirects	301/302 redirect chains	Chains (A→B→C); redirect loops	Medium

Core Web Vitals: Diagnosing & Fixing

LCP (Largest Contentful Paint) — Common Causes & Fixes

- Slow server response → Upgrade hosting, implement CDN, enable caching
- Render-blocking resources → Defer non-critical JavaScript; preload critical CSS
- Unoptimized hero images → Compress, use WebP, lazy-load below the fold
- Client-side rendering → Consider SSR for content-heavy pages

CLS (Cumulative Layout Shift) — Common Causes & Fixes

- Images without dimensions → Always specify width/height
- Ads or iframes without reserved space → Set explicit container sizes
- Dynamically injected content → Avoid inserting above existing content
- Web fonts causing FOUT → Use font-display: swap; preload key fonts

CASE STUDY 1

Medi-Station Urgent Care

From Mid-Tier Rankings to Map Pack Dominance

Client	Location	Industry	Campaign
Medi-Station Urgent Care	Miami Shores, FL	Healthcare / Urgent Care	Full-Stack Local SEO & GEO

The Challenge

Miami-Dade County has one of the highest densities of urgent care providers in the United States. In a market where a potential patient searches "urgent care near me" and sees ten options within a mile, visibility is not enough. Dominance is the only meaningful outcome.

When the campaign began, Medi-Station had inconsistent Google Map visibility, mid-tier organic rankings (positions 5–7 for core service keywords), limited presence in high-intent "near me" searches, underutilized local SEO infrastructure, and an underdeveloped Google Business Profile.

The System Applied

- Entity Validation & Optimization — Strengthened GBP signals; standardized NAP across 60+ directory listings.
- Off-Page Authority Expansion — Built citations on high-authority local and national health directories.
- Keyword Intent Alignment — Targeted high-intent "near me" searches with location-specific content.
- Conversion-Focused Experience — Improved user journey from search to action.
- Continuous Measurement & Optimization — Weekly ranking monitoring; data-driven adjustments.

The Results

Metric	Before	After
Average Google Position	5–7	Top 2–3
Google Map Pack Visibility	Inconsistent	Consistent first-page placement
"Urgent care near me" ranking	Outside top 3	Consistent top 3

Metric	Before	After
Geographic query coverage	Limited	Expanded across Miami-Dade
Lead quality	Mid-tier	Conversion-ready patient inquiries

"When the system is applied correctly — visibility increases, authority strengthens, conversions improve, and growth becomes predictable."

— Case Study 1 Takeaway

CASE STUDY 2

Ziemer Law, LLC — Evansville, Indiana

Scaling Visibility in a Competitive Legal Market

Legal services represent one of the most competitive and high-stakes SEO environments in digital marketing. The cost-per-click for personal injury keywords in Indiana routinely exceeds $50–$100. Organic rankings are fought over by large regional firms with dedicated marketing budgets and established domain authority.

Ziemer Law entered the campaign with an average Google position of 6 across core service keywords, limited Map Pack visibility for high-value legal searches, and underutilized GEO optimization for legal FAQ and service content.

The System Applied

- Entity Authority Building — Strengthened GBP; built citations on Avvo, FindLaw, Justia, Martindale-Hubbell.
- Competitive Keyword Strategy — Mapped high-volume, high-intent legal keywords in the Evansville market.
- GEO Content Optimization — FAQ-structured content; AI-extractable answers to top legal questions.
- Local Map Pack Optimization — Increased review velocity and recency on GBP.

- Continuous Monitoring & Refinement — Weekly tracking; competitor movement analysis.

The Results

Metric	Before	After
Average Google Position	6	60%+ improvement
Map Pack Visibility	Inconsistent	Consistent first-page placement
High-value legal keyword rankings	Page 2	Page 1
"Near me" legal searches	Limited presence	Strong local presence
Brand authority signals	Moderate	Strengthened across legal directories

Key Insight

Two industries. Two markets. Two completely different competitive environments. One system. Consistent results. The system does not depend on industry, niche, or market type. It depends on structure, data, and execution.

CASE STUDY 3

Goshen Village Trustee Election

How Paid Social Delivered a 3× Voter Turnout Surge

37,016 Impressions	**9,717** Reach	**1,125** Link Clicks	**$1,036.51** Total Spend	**$0.54** Cost Per Click	**6 Days** Campaign

The Challenge

Local elections rarely attract digital marketing expertise. In Goshen, New York, Village Trustee races had historically drawn around 335 voters. Christy and Jake Kriney — lifelong Goshen residents, parents, and local business owners — entered the March 18th election with a clear platform but limited time and a modest budget. The Meta campaign ran for just six days.

The Strategy

- Hyper-Local Geographic Targeting — Targeted only the Village of Goshen.
- Video-First Authentic Creative — Each candidate recorded their own video addressing one specific issue.
- Dedicated Landing Page — results4goshen.com gave undecided voters a clear, trustworthy destination.

Election Results

Voter Turnout: previous average 335 voters → 1,002 voters (+199%)

Christy Kriney: 285 votes — ELECTED

Jake Kriney: 247 votes — ELECTED

Value of the Win

Each elected trustee receives $9,000/year + full medical benefits. Total campaign cost: $1,036.51. This is the ROI of a system built on precise targeting, authentic creative, and a conversion-optimized landing page.

PART 3

SEARCH ENGINE MARKETING

Capturing Demand at the Moment It Matters

SEM is where intent becomes action. Part 3 covers the complete SEM system — landing page architecture, the 10-Y Framework, product pages, retargeting, lead generation, and competitive intelligence.

Search Engine Marketing (SEM)

Capturing and Scaling Demand at the Moment It Matters

"Nothing except the mint can make money without advertising."

— Thomas Babington Macaulay

If SEO and GEO build visibility and trust, SEM captures demand at the moment it matters most. Modern SEM is no longer manual — it is system-driven.

The Three Core Inputs of Modern SEM

1	2	3
Signal Quality	**Landing Page Experience**	**Campaign Structure**
How accurate and complete your conversion data is. The platform learns from what you track.	How well your page satisfies intent and converts users.	How clearly your campaigns communicate intent to the AI bidding system.

SECTION 3.1

Landing Page Architecture

Where Conversions Happen

In earlier SEM models, ad copy drove performance. Today, landing pages do. Platforms evaluate page speed, content relevance, engagement behavior, and conversion activity. Your ad gets you into the auction. Your landing page determines whether you stay competitive.

The Five Layers of High-Performing Landing Pages

1. Intent Alignment

The page must match exactly what the user searched for. Mismatch causes immediate drop in conversions.

2. Speed and Responsiveness

Load within 2.5s, respond instantly, no layout shifts. Affects user behavior and Quality Scores.

3. Above-the-Fold Clarity

Within the first screen: what you offer, why it matters, what to do next. Confusion kills conversions.

4. Trust Signals

Reviews, testimonials, certifications, guarantees. Trust reduces hesitation at the moment of decision.

5. Conversion Path

The next step must be obvious, easy, and fast. Short forms, click-to-call, or one-click checkout.

> **Continuous Testing: The Real Growth Lever**
>
> Landing pages are never finished. They are tested continuously. Focus on: headlines, CTA placement, form length, trust indicators, layout spacing.
>
> A 1% improvement in conversion rate on 1,000 monthly visits = 10 extra conversions. At an average deal value of $500, that is $5,000 in additional monthly revenue — from a single headline test.

SECTION 3.2

The 10-Step SEM Framework

The 10-Y System: Structured Execution for Modern Paid Search

1	2	3	4	5
QUANTIFY	**VERIFY**	**SOLIDIFY**	**NULLIFY**	**SPY**
Define CPA, ROAS, LTV, break-even before spending	Confirm tracking: GA4, Tag Manager, CRM, consent compliance	Build campaigns by intent clusters and funnel stages	Remove irrelevant traffic: negatives, search term audits	Analyze competitor ads, landing pages, messaging

6 AMPLIFY	7 DIVERSIFY	8 INTENSIFY	9 SIMPLIFY	10 CONTINUITY
Expand across Search, Display, YouTube, Demand Gen, PMax	Mix campaign types: Search, PMax, Retargeting	Scale 15–20% budget increments; protect machine learning	Remove underperformers; consolidate and clean structure	Report CPA, ROAS, conversion quality, revenue impact

Figure 3: The 10-Y Pay Per Click Management Framework

Quantify · Verify · Solidify · Nullify · Spy · Amplify · Diversify · Intensify · Simplify · Continuity

Step-by-Step Deep Dive

Step 1 — QUANTIFY: Define Success Before Spending

Target CPA, target ROAS, customer LTV, break-even thresholds. Without clear targets, optimization has no direction.

Step 2 — VERIFY: Infrastructure Before Investment

Ensure tracking is correct: GA4, Tag Manager, cross-domain tracking, conversion tracking, CRM integration, consent compliance.

Step 3 — SOLIDIFY: Campaign Architecture

Build campaigns based on intent clusters and funnel stages. Separate brand, service/product, geographic, competitor, and remarketing campaigns.

Step 4 — NULLIFY: Traffic Quality Over Volume

Remove irrelevant traffic through negative keywords, search term analysis, and ongoing weekly filtering. Quality outperforms volume.

Step 5 — SPY: Competitive Intelligence

Understand competitor ad copy, landing pages, offers, and keywords. Position strategically, not reactively.

Step 6 — AMPLIFY: Multi-Surface Expansion

Expand reach across Google Search, Display, YouTube, Demand Gen, and Performance Max.

Step 7 — DIVERSIFY: Campaign Type Mix

Search for high-intent capture, PMax for scaling, Demand Gen for mid-funnel, Display for remarketing.

Step 8 — INTENSIFY: Controlled Scaling

Scale gradually: 15–20% budget increments, expand audiences incrementally, test before scaling.

Step 9 — SIMPLIFY: Reduce to What Works

Periodically review structure: remove underperformers, consolidate where it improves signal volume, clean up structure.

Step 10 — CONTINUITY: Close-Loop Reporting

Measure CPA against target, ROAS against target, conversion quality (are leads closing?), and revenue impact.

SECTION 3.3

Product Pages

Where Paid Traffic Wins or Loses

For ecommerce businesses, product pages function as landing pages, conversion points, and SEO assets simultaneously. They must satisfy users, search engines, and AI systems.

Figure 4: The 8-Step Product Page Checklist

Fast Page Load · Tracking Code · Indexable · Quick Loading Media · Product Placement · Visible Add to Cart · Social Proof · Related Items

1	**Fast Performance**	Load under 2.5 seconds, stable, mobile-responsive

2	**Tracking Accuracy**	Accurate event tracking: add-to-cart, checkout initiation, purchase
3	**Indexability**	Search engines must understand and access every product page
4	**Media Optimization**	High-quality images (multiple angles), video demonstrations
5	**Product Placement**	Immediate understanding of what the product is and why it matters
6	**Strong CTA**	Clear, prominent Add to Cart or Buy Now with minimal friction
7	**Social Proof**	Reviews, star ratings, user-generated content, Q&A
8	**Recommendations**	Upsells, cross-sells, bundles, personalized recommendations

SECTION 3.4

The 90-Day SEM Scaling System

Scaling must be structured. This model provides a controlled, systematic approach to growing SEM performance over a 90-day period.

PHASE 1 — FOUNDATION (Days 1–30)

Focus: Clean data, stable performance, initial machine learning signal
Actions: Verify tracking, launch campaigns, collect clean data, remove waste
Goal: Stabilize CPA within 150% of target
Key Milestone: 30–50 conversion events per month — minimum for Smart

Bidding to optimize effectively

PHASE 2 — CONTROLLED SCALING (Days 31–60)

Focus: Gradual expansion while maintaining efficiency

Actions: Increase budgets 15–20% increments, expand audiences, improve creative

Goal: Increase conversion volume without degrading CPA

Key Milestone: CPA within 120% of target; ROAS above break-even

PHASE 3 — AGGRESSIVE EXPANSION (Days 61–90)

Focus: Maximizing volume while protecting profitability

Actions: Scale top performers, expand reach, optimize conversion rates

Goal: Build predictable, scalable growth

Key Milestone: CPA at or below target; ROAS above goal

Scaling Rules

Scale when: CPA stable for 2+ consecutive weeks; 30+ conversions/month per campaign; tracking confirmed accurate.

Pause scaling when: CPA spikes >30% above target; conversion rate drops >20%; data becomes inconsistent.

The Scaling Philosophy

Do not scale traffic. Scale: conversion rate, signal quality, audience quality, and system clarity. When these improve, scaling becomes safe. When they are weak, scaling amplifies the problem.

SECTION 3.5

Google Ads Creative & Ad Copy Framework

Writing Ads That Win the Click and Earn the Conversion

In the AI-driven SEM environment of 2026, platforms handle bidding, audience expansion, and placement decisions automatically. But there is one element the platform cannot generate: compelling, differentiated ad copy that earns the click.

Responsive Search Ads (RSA): The Modern Ad Structure

Element	Count	Character Limit	Strategy
Headlines	Up to 15 (min 8)	30 chars	Cover intent, benefits, CTAs, differentiation
Descriptions	Up to 4 (min 3)	90 chars	Expand on headlines; include social proof and urgency
Display URL	2 path fields	15 chars	Reinforce relevance with keyword-aligned paths
Ad Strength	Google score	Excellent	Target Excellent — more diverse, relevant assets

The 5 Headline Categories Every RSA Needs

1	2	3	4	5
Intent Match	**Value Proposition**	**CTA**	**Trust & Proof**	**Urgency / Scarcity**
Mirror the search query directly	Communicate the primary benefit	Tell the user exactly what to do	Build credibility, reduce risk	Create a reason to act now

Quality Score: The Hidden Cost Multiplier

Quality Score (1–10) directly determines your CPC and ad position. A QS of 10 can result in CPCs 50% lower than a QS of 5 — for the exact same position.

Quality Score Component	Weight	How to Improve
Expected Click-	~35%	Compelling, relevant headlines that

Quality Score Component	Weight	How to Improve
Through Rate		match query intent
Ad Relevance	~35%	Align ad copy with specific keyword ad groups
Landing Page Experience	~30%	Fast load, relevant content, clear CTA, low bounce rate

SECTION 3.6

Retargeting Architecture

Converting the 97% Who Didn't Act on First Visit

On average, 97% of first-time website visitors leave without converting. Retargeting is the system that re-engages those visitors — showing them relevant ads as they browse other sites, watch YouTube, check social media, or search Google again.

The Full-Funnel Retargeting Sequence

Timeline	Stage	Message Focus	Objective
Day 1–3	Awareness	Brand content — who you are, what you do	Stay top of mind
Day 4–7	Considerati on	Testimonials, case studies, comparisons	Build trust and differentiation
Day 8–14	Offer	Specific offer, discount, free consultation	Drive conversion
Day 15–30	Final	Last-chance messaging, social proof	Capture fence-sitters
Day 30+	Re-engageme nt	Seasonal offers, new content	Reactivate dormant prospects

Retargeting Audience Segments

Audience Segment	Strategy	Intent
All Website Visitors (30 days)	Brand awareness retargeting	Low
Specific Page Visitors	Targeted offer ads	Medium
Pricing Page Visitors	Urgency and value messaging	High
Cart Abandoners (ecommerce)	Direct product recovery ads	Very High
Lead Form Starters	Simplicity messaging	Very High
Video Viewers (25%+)	Graduated intent retargeting	Medium–High
Customer Match Lists	Upsell, cross-sell, reactivate	Highest

The Retargeting ROI Principle

Retargeting users convert at 3–10× the rate of cold audiences. The cost difference is minimal. The revenue difference is enormous. A retargeting campaign with a $500/month budget consistently outperforms a $5,000/month cold traffic campaign for businesses with established website traffic.

SECTION 3.7

Lead Generation Systems

Designing the Full Journey from Click to Qualified Lead

Generating clicks is not the goal of SEM. Generating qualified leads that close is the goal. A lead generation system is the structured sequence of touchpoints, pages, and follow-up actions that converts a search query into a real business opportunity.

The 5-Minute Lead Response Rule

Studies consistently show that responding to a web lead within 5 minutes increases the likelihood of qualifying that lead by 21×

compared to responding within 30 minutes. After 5 minutes, that probability drops sharply with every passing minute. The best marketing in the world is wasted if the lead sits in an inbox for 4 hours.

Lead Generation Funnel Architecture

1	**1. Traffic Source**	Qualified click from Google Ads, organic search, or Paid Social
2	**2. Landing Page**	Intent-matched page presenting a clear, compelling reason to take action
3	**3. Conversion Action**	Phone call (highest intent), short form (3–4 fields), chat, or appointment booking
4	**4. Immediate Follow-Up**	Automated confirmation within 60 seconds; phone callback within 5 minutes
5	**5. Lead Qualification**	First conversation: qualify need, timeline, budget, decision authority
6	**6. Nurture**	Multi-touch sequence: email + retargeting + SMS over 7–30 days
7	**7. Closed Won/Lost**	Revenue attribution back to original source; data feeds optimization

Form Optimization: The Conversion Rate Lever

Form Element	Best Practice	Why
Number of fields	3–5 fields max for initial capture	Each extra field reduces completions ~10%

Form Element	Best Practice	Why
Required fields	Name, Phone, Service needed	Phone enables fast follow-up
CTA button text	"Get My Free Quote" not "Submit"	Action-oriented communicates value
Form headline	Repeat the value proposition	Reinforces why to complete
Social proof	Star rating + review count near CTA	Reduces hesitation at decision
Privacy statement	1-line disclaimer below submit	Reduces spam concern
Mobile optimization	Large fields, auto-fill enabled	60%+ of form fills are mobile

SECTION 3.8

Competitive Intelligence Framework

Understanding Your Competitive Landscape to Win Your Market

In every market, there are businesses that spend less and get better results — not because they found a magic tactic, but because they understand their competitive landscape better than their competitors do. Important: the goal is not to copy. It is to understand what is working in your market and why — then do it better, or do it differently.

The 5 Domains of Competitive Intelligence

1. Search Visibility

Which keywords do competitors rank for organically and bid on in paid search?

Tools: Semrush, Ahrefs, Google Ads Transparency Center

2. Content Strategy

What content do competitors publish, how often, and what gets the most links?

Tools: Ahrefs Content Explorer, BuzzSumo, manual review

3. Ad Creative & Messaging

What ads run? What offers and landing pages? How is value framed?

Tools: Google Ads Transparency Center, Meta Ad Library, SpyFu

4. Reputation & Reviews

What do customers say about competitors? What complaints repeat?

Tools: Google Reviews, Yelp, industry review platforms

5. Backlink Profile

Who links to competitors that does not link to you? What earned the most links?

Tools: Ahrefs, Semrush Link Building Tool, Majestic

Ongoing Competitive Monitoring

- Google Alerts for competitor brand names and key personnel
- Semrush/Ahrefs position tracking for shared keyword sets
- Monthly review of competitor Google Business Profile reviews
- Quarterly review of competitor ad creative and landing pages

PART 4

PAID SOCIAL

Demand Creation & Creative Systems

If SEM captures existing demand, Paid Social creates it. Part 4 covers the Creative-First Framework, the 10-T System, full-funnel strategy, and the 90-Day Paid Social Scaling System.

Paid Social: Demand Creation

Creative-First Strategy for Modern Social Advertising

If SEM captures existing demand, Paid Social creates it. Paid Social introduces your brand to people who are not actively searching, may not know your solution exists, and have not yet entered the buying cycle.

Search is reactive. Paid Social is proactive. This fundamental difference changes everything about how you build campaigns, what you measure, and how you optimize.

The New Paid Social Performance Equation

1	2	3
Creative Quality	**Signal Strength**	**Testing Velocity**
How engaging your content is — the platform amplifies what works	How accurate your conversion data is — bad signals = bad optimization	How quickly you iterate — the business that tests fastest wins

SECTION 4.1

The Creative-First Framework

"Creativity is contagious. Pass it on."

— Albert Einstein

Every piece of creative must accomplish three things:

1. Capture Attention Immediately

You have 1–2 seconds on a feed. If you fail here, nothing else matters. The hook — visual or verbal — must stop the scroll before anything else can work.

2. Maintain Attention

The user must keep watching, reading, or engaging through the full message. Fast-paced structure, relatable problems, and native platform feel all serve this goal.

3. Drive Action

The user must know exactly what to do next. Clear, specific calls-to-action are not optional. Vague CTAs produce vague results.

High-Performing Creative Types

Creative Type	Why It Works	Best For
User-Generated Content (UGC)	Authentic, trusted, platform-native	All verticals — especially local
Founder-Led Videos	Personal authority, direct trust	Professional services, B2B
Customer Testimonials	Social proof from real users	High-consideration purchases
Before-and-After	Visual proof of outcome	Home services, health, fitness
Problem/Solution	Creates recognition and desire	Any product solving a clear pain
Product Demonstrations	Shows the product solving a problem	Ecommerce, SaaS, tech
Comparison Content	Positions clearly against alternatives	Competitive markets

Creative Insight

85% of social video is watched without sound. Always include subtitles. A smartphone video from a real customer outperforms polished studio production in most verticals.

Creative Fatigue: The Silent Performance Killer

Even great creative stops working over time. Signs of fatigue: rising CPA, increasing frequency, declining CTR, reduced engagement.

Solution: continuous creative refresh. New hooks, new angles, new formats — every two to four weeks. The businesses that win in Paid Social are not those with the best single creative. They are those with the most consistent testing pipeline.

SECTION 4.2

The 10-Step Paid Social Framework

The 10-T System

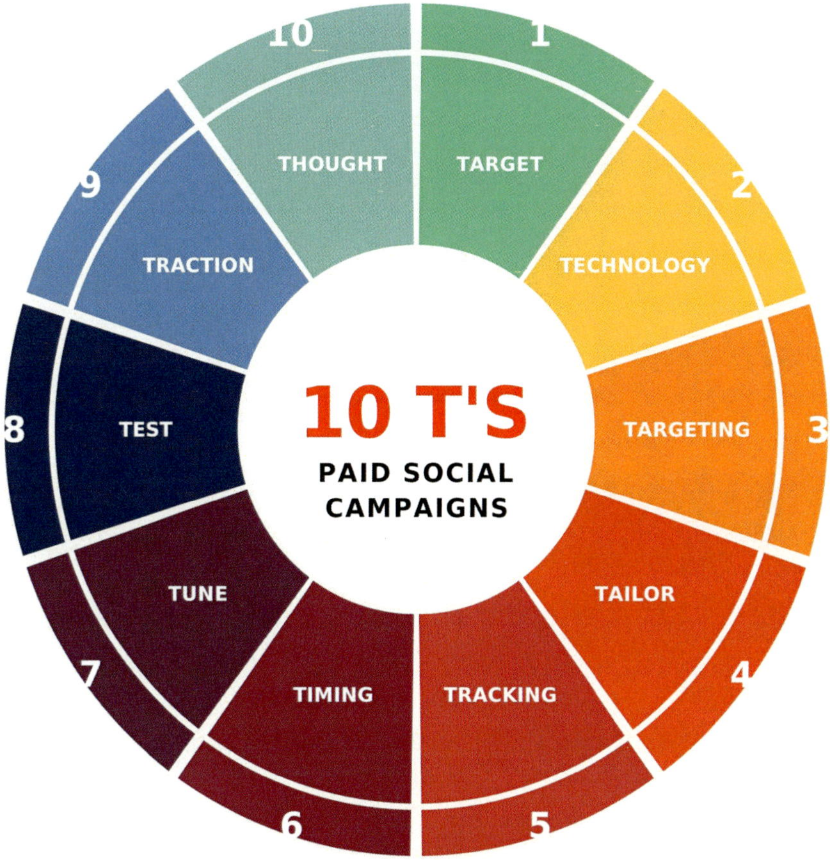

Figure 5: The 10-T Paid Social Campaigns Framework

Target · Technology · Targeting · Tailor · Tracking · Timing · Tune · Test · Traction · Thought

Step-by-Step Deep Dive

Step 1 — TARGET OUTCOME: Define the Goal

AI needs a clear optimization goal. Define CPA, ROAS, CPL, LTV. These are not reporting metrics — they are the targets the platform uses to optimize delivery.

Step 2 — TECH STACK: Infrastructure Before Launch

Before spending: Meta Pixel verified, CAPI implemented server-side, GA4 connected, CRM integration active, server-side tracking configured. CAPI is no longer optional in 2026.

Step 3 — TARGET SIGNALS: Trust the AI

Use broad targeting and feed the platform high-quality conversion data. Over-targeting reduces scale and prevents the algorithm from finding your best audience.

Step 4 — THUMB-STOP CREATIVE: Win the First Second

Strong hooks (visual or verbal in first second), vertical formats (9:16), native content styles, fast engagement.

Step 5 — TRACK PERFORMANCE: The Right Metrics

Cost per result, CTR, watch time, conversion rate, ROAS, frequency. Always tie ad performance to actual revenue outcomes.

Step 6 — TEST VELOCITY: Volume Wins

New creatives every week, multiple hooks, multiple angles. The testing pipeline is the competitive advantage — not any single creative.

Step 7 — TWEAK STRATEGICALLY: Kill or Scale

Kill underperformers; scale winners. Avoid constant micro-changes that prevent the algorithm from stabilizing.

Step 8 — TIER BUDGETS: Allocate Strategically

Allocate across scaling campaigns, testing campaigns, and experimental campaigns.

Step 9 — TRACK INCREMENTALITY: Beyond Platform Metrics

Measure branded search lift, organic traffic growth, and direct traffic increases — to understand the full impact of Paid Social.

Step 10 — TREND FORWARD: Stay Current

Stay aligned with platform updates, creative trends, and new formats. The format that worked last quarter rarely works next quarter.

SECTION 4.3

Full-Funnel Paid Social Strategy

Users are at different stages of awareness. Your campaigns must match each stage. A user who has never heard of your brand needs a different message than a user who visited your website yesterday.

TOP OF FUNNEL — Awareness
Goal: Introduce your brand and capture attention

Audience: Broad, cold audiences; interest-based or lookalike
Creative: Storytelling, problem awareness, educational
KPI: CPM, video views, reach
Budget: 30–40% of budget

MIDDLE OF FUNNEL — Consideration

Goal: Build trust and show differentiation
Audience: Video viewers, website visitors, social engagers
Creative: Testimonials, case studies, comparisons
KPI: CPC, landing page views
Budget: 25–35% of budget

BOTTOM OF FUNNEL — Conversion

Goal: Drive direct action
Audience: High-intent visitors, cart abandoners, lead form openers
Creative: Specific offers, urgency, social proof
KPI: CPA, ROAS, conversion rate
Budget: 30–40% of budget

The Retargeting Layer

Retargeting targets users who visited your site (30–180 days), watched 50%+ of your video ads, engaged with your social profiles, or interacted with previous ads but did not convert. Users who already know your brand convert at 3–5× the rate of cold audiences. Retargeting is not optional — it is a revenue layer that belongs in every paid social system.

SECTION 4.4

The 90-Day Paid Social Scaling System

PHASE 1 — FOUNDATION (Days 1–30)

Focus: Clean tracking, creative testing, data collection
Actions: Launch 4–6 creatives per ad set, identify winners, remove underperformers
Goal: Stabilize cost per result
Creative Output: 8–12 creative variations tested

PHASE 2 — STRUCTURED SCALING (Days 31–60)

Focus: Increasing spend while maintaining efficiency
Actions: Increase budgets 20%/week on winners, duplicate winning ad sets, expand targeting
Goal: Grow volume while maintaining efficiency
Creative Output: New creatives weekly; fresh angles on winning hooks

PHASE 3 — AGGRESSIVE GROWTH (Days 61–90)

Focus: Maximizing reach and improving conversion efficiency
Actions: Increase creative production, scale best campaigns, optimize landing pages
Goal: Build predictable, scalable performance
Creative Output: 3–5 net new creatives per week

> **The Paid Social Growth Principle**
>
> Paid Social is not controlled by targeting. It is controlled by creative, data, and iteration. When those improve, performance scales. When they stagnate, performance plateaus.

SECTION 4.5

Video Marketing & Creative Production

The Dominant Format of Modern Digital Marketing

Video accounts for over 80% of all internet traffic by 2026. Platforms algorithmically reward video content with greater organic reach, lower advertising costs, and higher engagement rates than any other format.

The Video Funnel Strategy

1	2	3
SHORT-FORM (15–60 sec)	**MID-FORM (60–180 sec)**	**LONG-FORM (3–20 min)**
TikTok, Reels, YouTube Shorts. Awareness & attention capture.	Instagram, Facebook, YouTube. Consideration & education.	YouTube, website. Deep trust & conversion.

The Hook Framework: Winning the First 3 Seconds

Hook Type	Example Opening	Why It Works
The Bold Claim	"Most [people] don't know this about [topic]..."	Creates curiosity through information gap
The Problem	"If you're struggling with [pain], this is for you."	Immediate audience qualification
The Counterintuitive	"Stop doing [common practice]. Here's what works."	Challenges assumption, demands attention
The Result	"I went from X to Y in [timeframe] — here's how."	Leads with outcome to create desire
The Question	"Have you ever wondered why [X happens]?"	Engages curiosity immediately
The Visual	Dramatic visual or unexpected scene in first frame.	Pattern interrupt — works on muted feeds

PART 5

REPUTATION, ANALYTICS & GROWTH

Where Everything Connects

Reputation determines whether traffic converts. Analytics determines whether growth is real. Part 5 ties every system together into the SEO AI Growth Engine — a compounding loop that produces predictable, scalable results.

SECTION 5.1

Reputation Management

Controlling How Your Brand Is Perceived

> *"It takes 20 years to build a reputation and five minutes to ruin it. If you think about that, you will do things differently."*
>
> — Warren Buffett

Reputation is no longer just reviews. It is your entire digital footprint — Google reviews, third-party platform ratings, social media conversations, news and media mentions, forum discussions, and AI-generated summaries. In 2026, AI systems summarize your reputation before users even click.

Figure 6: The 8-Step Reputation Management Framework

Regroup · Review · Request · Repair · Respond · Release · Recycle · Refine

The 8-Step Reputation Management Framework

1	REGROUP	**Define Your Narrative** Define your mission, positioning, and value proposition. If you do not define your narrative, the internet will.

2	REVIEW	**Audit Your Current Status** Search your brand and analyze reviews, AI summaries, mentions. What patterns emerge? This is your reputation brief.

3	REQUEST	**Build a Review Generation System**

		Automated email/SMS, CRM-triggered flows, QR codes. Focus on consistency, authenticity, detail.
4	REPAIR	**Fix What's Broken** Duplicate listings, incorrect business info, fake/spam reviews. Consistency strengthens trust.
5	RESPOND	**Every Review Gets an Answer** Within 24–48 hours. Professional, empathetic. Your response matters more than the review itself.
6	RELEASE	**Activate Reviews as Marketing** Use across website, paid ads, social, email. Reputation should influence conversion at every touchpoint.
7	RECYCLE	**Turn Reviews Into Creative** Screenshot UGC ads, social quote graphics, landing page pull-quotes. Customer voice carries more credibility.
8	REFINE	**Ongoing Monitoring** Sentiment trends, brand mentions, AI summary accuracy. Reputation is not a one-time task.

SECTION 5.2

Analytics: From Data to Decision-Making

Without analytics, you guess, misallocate budget, and scale the wrong things. The moment a decision is made based on month-old data, that decision is already 30 days behind reality. In a fast-moving digital environment, 30 days is an eternity.

Building the Analytics Dashboard

The executive dashboard should answer four questions at a glance:

1	2	3	4
How much did we spend?	How much did we generate?	What is our acquisition cost?	What is our return?

Everything else supports these four questions. A dashboard with 40 metrics and no clear answers to these four questions is not an analytics system. It is a library.

Recommended Tools for the Unified Workspace

Tool	Function	Priority
Google Analytics 4 (GA4)	Web behavior, conversion tracking, attribution	Required
Google Search Console	Organic search, indexing, Core Web Vitals	Required
Looker Studio	Unified dashboards across all data sources	Required
Google Ads Dashboard	Paid search performance: CPA, ROAS, QS	Required
Meta Ads Manager	Paid social performance: CPR, ROAS, creative	Required
GoHighLevel / HubSpot	CRM pipeline, close rate, revenue by source	Required
CallRail / WhatConverts	Call tracking and offline conversion attribution	Recommended

Stop reporting on activity. Start reporting on outcomes.

SECTION 5.3

The SEO AI Growth Engine

How Every System Feeds Every Other System

At this point, every system you have built feeds the others. SEO builds organic visibility. GEO ensures AI understanding and citation. Off-page SEO builds domain authority and brand recognition. SEM captures high-intent demand. Paid Social creates brand demand. Reputation builds trust. Analytics measures everything and directs optimization.

The Compounding Growth Loop

1 **Visibility** → Traffic	**2** **Traffic** → Conversions	**3** **Conversions** → Customers
4 **Customers** → Reviews	**5** **Reviews** → Trust	**6** **Trust** → Higher CVR
7 **Higher CVR** → Better Data	**8** **Better Data** → AI Optimization	**9** **Optimization** → Efficiency

Each iteration of the loop is stronger than the last.

Why Most Businesses Fail to Scale

Not because they lack tactics. Because:

1 **Systems Are Disconnected** Channels operate in silos. No compounding effect. Each channel optimizes for its own metrics without visibility into how it influences the others.	**2** **Data Is Incomplete** Decisions are made on partial information. Wrong bets are scaled. Winning bets are cut. The difference between growing and stagnating often comes down to data quality.	**3** **Execution Is Inconsistent** Tactics are tried and abandoned before compounding begins. Systems require sustained execution — not bursts of activity followed by inaction.

The Final Growth Principle

What you measure together is what you grow together. When data is unified, patterns become visible, waste becomes obvious, and opportunities become scalable. Marketing and sales alignment is not a culture issue. It is a data infrastructure issue.

SECTION 5.4

Email Marketing & CRM Automation

Owning the Channel: Building Revenue-Generating Relationships

Every other channel in this book — search, paid social, video, reputation — drives traffic that you do not own. When a platform changes its algorithm, raises ad prices, or shuts down, that traffic disappears. Email is the one channel where you own the audience.

An email list of 10,000 qualified subscribers with a 25% open rate reaches 2,500 engaged people for $0 in additional ad spend. The economics of email are unmatched: industry average email ROI is $36 for every $1 spent.

The Email Marketing System Architecture

1	LIST BUILDING	Lead magnets, opt-in forms, landing page capture, CRM integration
2	SEGMENTATION	By source, behavior, stage, interest, and customer lifetime value
3	AUTOMATION FLOWS	Welcome, nurture, re-engagement, post-purchase, review request
4	BROADCAST CAMPAIGNS	Newsletters, promotions, announcements, content shares
5	ANALYTICS & OPTIMIZATION	Open rate, CTR, conversion, unsubscribe, revenue attribution

Essential Automation Sequences

Welcome Sequence (Days 1–7)

- Email 1 (Immediate): Deliver lead magnet + introduce brand story
- Email 2 (Day 2): Share your most valuable piece of content
- Email 3 (Day 4): Social proof — case study or success story
- Email 4 (Day 7): Soft offer — free consultation, trial, or next step

Lead Nurture Sequence (Weeks 2–8)

- Weekly educational content relevant to their stated interest
- 3:1 ratio — three value emails per one offer email
- Progressive trust building — each email answers a common objection
- Behavior triggers — pricing-page click triggers a more direct follow-up

> **CRM as Competitive Advantage**
>
> The businesses that scale most efficiently are those that know, with certainty, which marketing dollar produces the most revenue. That certainty requires a CRM that captures source data at the point of lead capture and tracks it through to closed revenue — not just to form fill.

SECTION 5.5

AI Tools & Marketing Automation

Operationalizing Artificial Intelligence Across Your Marketing System

AI has moved from an experimental add-on to a core operational layer. The goal is not to use AI for its own sake — it is to produce better outputs, faster, with less manual effort, freeing your team for strategy, relationships, and decisions that require human judgment.

The AI Marketing Stack by Function

Function	AI Application	Recommended Tools
Content Creation	Articles, blog posts, emails, social captions	Claude, ChatGPT, Jasper
SEO Research	Keyword clustering, competitor analysis	Semrush AI, Ahrefs AI
Ad Copy	Headline variations, RSA assets	Claude API, Google Ads AI
Image & Video Creation	Ad creative, social graphics, product images	Midjourney, Adobe Firefly, Runway
Analytics & Reporting	Data interpretation, anomaly detection	GA4 AI insights, Looker AI
Customer Service	FAQ chatbots, lead qualification	GoHighLevel AI, Intercom AI
Email Personalization	Subject line, send time, content blocks	Klaviyo AI, Mailchimp AI
Bid Management	Real-time bid optimization across campaigns	Google Smart Bidding, Meta Advantage+

The Marketing AI Prompt Framework

The quality of AI output is determined almost entirely by the quality of the input. Every effective marketing AI prompt has four elements:

1	2	3	4
ROLE	**CONTEXT**	**TASK**	**FORMAT**
Define who the AI should act as	Specific business and audience details	The specific, clear instruction	Specify the exact output format needed

The Human-AI Partnership Principle

AI does not replace the strategist. It replaces the task. The strategic value of a marketing team is not in writing first drafts, compiling data, or generating variations — it is in knowing which draft to run, which data matters, and which variation to test first. AI returns that strategic time to the humans who should be using it.

SECTION 5.6

The Client Reporting & Communication System

Turning Data Into Clarity, Confidence, and Alignment

Reporting is not about proving you did work. It is about translating performance data into business decisions, maintaining alignment with stakeholders, and building the trust that enables long-term strategic relationships.

The Reporting Cadence Framework

1	**Real-Time Dashboard**	Always on — campaign performance, spend, leads, ROAS visible at any moment	Marketing team, business owner
2	**Weekly Performance Review**	Every Monday — prior week vs. goals, top wins, issues, actions	Marketing team + client point of contact
3	**Monthly Strategy Report**	1st of each month — full channel breakdown, trends, budget, goal tracking, plan	Business owner + leadership
4	**Quarterly Business Review**	Every 90 days — strategic review, ROI calc, next quarter roadmap	Full leadership team

Common Reporting Mistakes to Avoid

- Reporting on activity instead of outcomes ("We published 8 blog posts" vs "Organic traffic grew 23%")
- Using vanity metrics — clicks, impressions, followers without conversion context

- Presenting data without interpretation — numbers without insight create confusion
- Monthly-only reporting — problems fester for 30 days before being identified
- Omitting bad news — declining performance that isn't reported isn't being managed
- Missing the revenue connection — every report should answer "what did this make us?"

SECTION 5.7

The Social Proof System

Turning Customer Outcomes Into Marketing Assets

Social proof is the most powerful conversion accelerator available to any business. It works because users trust other users more than they trust businesses. The businesses that systematize social proof collection and distribution — rather than treating it as an ad-hoc activity — create a compounding trust asset that improves conversion rates across every channel simultaneously.

The Social Proof Hierarchy

1	**Case Studies**	Most powerful — specific results, named clients (with permission)	Strongest
2	**Video Testimonials**	Real person, real voice, real emotion	Very strong
3	**Written Testimonials**	High value when specific and detailed	Strong
4	**Star Ratings & Counts**	Visible in SERPs, Maps, and ads. Critical for click-through	Strong

5	**Media Mentions & Awards**	Third-party validation from recognized sources	Moderate
6	**User-Generated Content**	Organic photos and videos from real customers	Moderate
7	**Social Follower Counts**	Weakest — counts without engagement are easy to manipulate	Weak

SECTION 5.8

Advanced Attribution & Revenue Tracking

Connecting Marketing Activity to Business Revenue

Most businesses know their marketing costs with precision. Few know their marketing returns with the same precision. Advanced attribution closes that gap — connecting every marketing dollar spent to every revenue dollar generated, through the full customer journey.

Attribution Models: Understanding the Options

Model	How It Works	Best Used When	Limitation
Last Click	100% credit to final touchpoint	Simple campaigns, single channel	Ignores prior touchpoints
First Click	100% credit to first touchpoint	Brand awareness measurement	Ignores conversion-driving touch
Linear	Equal credit across all touchpoints	Multi-channel balanced campaigns	Doesn't reflect importance
Time Decay	More credit to recent touchpoints	Short sales cycles	Undervalues brand-

Model	How It Works	Best Used When	Limitation
			building
Position-Based	40% first, 40% last, 20% middle	Clear acquisition + conversion stages	Middle still undervalued
Data-Driven	AI assigns credit based on actual contribution	1,000+ conversions/month	Requires significant data

The Revenue Attribution Stack

True attribution requires connecting data across multiple systems. Build the stack in order:

1	**Layer 1 — UTM Parameters**	Tag every URL with source, medium, campaign, content. Foundation — without UTMs, attribution is guesswork.
2	**Layer 2 — GA4 Event Tracking**	Track micro-conversions (page views, scrolls, clicks) and macro-conversions (form fills, calls, purchases).
3	**Layer 3 — Call Tracking**	Dynamic tracking numbers per traffic source. CallRail or WhatConverts records which ad/keyword drove each call.
4	**Layer 4 — CRM Source Tracking**	Capture UTM data from first visit and store it in the CRM record. Connects ad to deal even 30 days later.
5	**Layer 5 — Revenue Attribution**	When a deal closes, tag revenue with original marketing source. Feed back to ad platforms as offline conversions.
6	**Layer 6 — Unified Dashboard**	GA4 + CRM + call tracking in Looker Studio. Single view: spend → leads → revenue → ROI by channel.

The Attribution Insight That Changes Everything

The most valuable attribution insight is rarely which channel generated the most leads. It is which channel generated the leads that actually closed. A channel generating 50 leads at a 40% close rate outperforms a channel generating 200 leads at a 5% close rate — at the same cost. You can only see this if your CRM is connected to your marketing data.

GROWTH FRAMEWORK

The Scaling Playbook

A Stage-by-Stage Framework for Building Market Dominance

Growth is not linear. Businesses that scale successfully move through distinct stages, each requiring a different strategic focus, different channel mix, and different success metrics.

The Four Stages of Marketing Scale

1	2	3	4
STAGE 1 Foundation	**STAGE 2 Momentum**	**STAGE 3 Scaling**	**STAGE 4 Dominance**
$2,000–$8,000/month Months 1–6 First 10 qualified leads/month at target CPA	$8,000–$20,000/month Months 7–18 Consistent 30+ leads/month; CPA below target	$20,000–$60,000/ month Months 19–36 Month-over-month revenue growth; ROAS above target	$60,000+/month Year 3+ Category leadership; inbound referrals; brand-search growth

The Decision Framework: When to Scale vs. When to Optimize

1	2	3
GREEN LIGHT Scale Now	**YELLOW LIGHT Optimize First**	**RED LIGHT Pause & Fix**
CPA at or below target for 3+ weeks. 30+ conversions/month. Tracking accurate. Conversion rate stable. Lead quality confirmed.	CPA within 25% above target but trending down. Insufficient conversion volume. Creative fatigue signals. Recent platform	CPA 50%+ above target for 7+ days. Tracking anomalies. Sales reporting lead-quality drops. Platform policy issues. Significant budget overspend.

changes.

GLOBAL ADDITION

The Modern Marketing Stack

Every system in this book depends on infrastructure. Tools do not create performance — systems do. But without the right tools, systems cannot function.

The Complete Stack by Layer

Visibility Layer	**Google Search Console, Ahrefs / Semrush**	Organic and AI search performance monitoring
Schema Layer	**Schema.org (schema.org)**	Structured data for machine understanding
Acquisition Layer	**Google Ads, Meta Ads Manager**	Paid demand capture and creation
Conversion Layer	**Website, landing pages, CRO tools**	Turning traffic into revenue
Data Layer	**GA4 (analytics.google.com), GTM**	Tracking and attribution
CRM Layer	**GoHighLevel, HubSpot**	Lead management and revenue tracking
Reputation Layer	**Google Business Profile (google.com/business)**	Local visibility and review management

Visualization Layer	**Looker Studio (lookerstudio.google.com)**	Unified dashboard and reporting
Call Tracking	**CallRail, WhatConverts**	Offline conversion attribution

The 6 Rules of the Marketing Operating System

1. Infrastructure Comes First

Before campaigns, before creative, before content — build the foundation.

2. Data Must Be Accurate

Platforms optimize based on what they are told. Wrong data = wrong performance.

3. Systems Must Be Connected

Advantage comes from compounding effect of connected channels.

4. Creative Must Evolve

The creative that worked last quarter will not work next quarter.

5. Trust Must Be Built

Before users convert, they validate. Trust cannot be purchased — it must be earned.

6. Optimization Must Be Continuous

The best system today is the system improved most consistently over time.

FINAL CHAPTER

The Operating System of Modern Marketing

Marketing is no longer a campaign, a tactic, or a channel. It is an operating system. The businesses that win do not win because they found a secret tactic. They win because they built a system and executed it with discipline, long enough for it to compound.

What Actually Wins

1	2	3
CLARITY	**CONSISTENCY**	**EXECUTION**
Knowing exactly what you want, what you measure, and what success looks like	Executing the plan with discipline, not reacting to every platform update	Building the system, connecting the parts, committing long enough to compound

The Path Forward

- Define your plan clearly — know your CPA, ROAS, and growth targets
- Build your infrastructure correctly — tracking, CRM, analytics before campaign spend
- Connect your systems — SEO, GEO, SEM, Paid Social, Reputation, Analytics share data
- Track performance accurately — unified analytics, real-time dashboards, revenue attribution
- Optimize continuously — weekly cycles, 90-day sprints, creative refresh pipelines
- Put your customer first — understand intent, reduce friction, build trust, deliver value

"Marketing is not about manipulating algorithms. It is about understanding people, solving problems, building trust, and measuring honestly. When you do that, platforms reward you, data confirms it, and growth follows."

— John Vincent Kriney

"The system is the advantage."

APPENDIX A

Glossary of Key Terms

Defined terms are presented with extra spacing for scannability.

A/B Testing

Running two versions of a page, ad, or creative simultaneously to determine which performs better based on data.

Attribution

The process of assigning credit for a conversion to the correct marketing touchpoints that influenced the decision.

Brand Mention

Any reference to a brand name across the web, whether linked or unlinked — both count as authority signals.

CAC (Customer Acquisition Cost)

The total cost to acquire one paying customer, including all marketing and sales expenses.

CAPI (Conversions API)

Meta's server-side tracking solution that sends conversion data directly from the server, bypassing browser-based limitations.

CLS (Cumulative Layout Shift)

A Core Web Vital measuring visual stability — how much page layout shifts unexpectedly during load.

Click-Through Rate (CTR)

The percentage of users who click on an ad or search result after seeing it.

Conversion Architecture

The structured design of a digital experience to guide users toward a specific action or outcome.

Conversion Rate

The percentage of visitors who complete a desired action (call, form fill, purchase).

Core Web Vitals

Google's set of user experience metrics: LCP, INP, and CLS — all measured against specific targets.

CPA (Cost Per Acquisition)

The cost to acquire one customer or conversion event.

CPM (Cost Per Thousand Impressions)

The cost to serve an ad to one thousand users — common in awareness campaigns.

CRM (Customer Relationship Management)

Software used to manage leads, customer interactions, pipeline stages, and revenue tracking.

Digital PR

The practice of earning media coverage, editorial mentions, and links from news and industry publications.

E-E-A-T

Experience, Expertise, Authoritativeness, Trustworthiness — Google's framework for evaluating content quality.

Entity

How Google understands a distinct real-world object (business, person, place) and its attributes and relationships.

First-Party Data

Data collected directly from users through your own platforms — forms, purchases, subscriptions, behavior.

Frequency

In Paid Social, the average number of times a user sees the same ad within a given period.

GA4 (Google Analytics 4)

Google's current analytics platform, event-based and designed for cross-platform measurement.

GEO (Generative Engine Optimization)

The practice of optimizing content for AI-powered summary and citation systems.

GTM (Google Tag Manager)

A tag management system that allows deployment and management of tracking codes without code changes.

INP (Interaction to Next Paint)

A Core Web Vital measuring page responsiveness — how quickly a page responds to interaction.

Intent

The underlying purpose behind a user's search query or behavior.

LCP (Largest Contentful Paint)

A Core Web Vital measuring how quickly the largest visible content element loads.

LTV (Lifetime Value)

The total revenue a customer generates over their entire relationship with a business.

Map Pack

The block of three local business results that appears in Google Search for local queries.

NAP

Name, Address, Phone Number — the three pieces of business data that must be consistent across all platforms.

Negative Keywords

Keywords you exclude from paid campaigns to prevent ads from showing for irrelevant searches.

Off-Page SEO

SEO signals that come from outside your website — links, mentions, reviews, citations.

On-Page SEO

SEO elements on your own website — content, structure, speed, metadata, structured data.

Paid Social

Advertising run on social media platforms such as Meta, TikTok, LinkedIn, YouTube.

Performance Max

Google's AI-driven campaign type that serves ads across all Google surfaces simultaneously.

Remarketing / Retargeting

Showing ads to users who have previously visited your site or interacted with your content.

ROAS (Return on Ad Spend)

Revenue generated divided by advertising cost — the fundamental paid media efficiency metric.

Schema Markup

Structured data added to your HTML to help search engines understand your content.

SEM (Search Engine Marketing)

Paid advertising on search engines, primarily Google Ads.

SEO (Search Engine Optimization)

The practice of optimizing websites to rank higher in organic search results.

SERP

Search Engine Results Page — the page Google displays in response to a search query.

Smart Bidding

Google's AI-driven automated bid strategies, optimizing toward your stated conversion goals.

Structured Data

Machine-readable code (JSON-LD schema) that tells search engines what your content means.

UGC (User-Generated Content)

Content created by real users — reviews, testimonials, videos — used as creative.

Unified Analytics

A single, integrated analytics system that connects all channel data into one source of truth.

APPENDIX B

Recommended Tools & Resources

Visibility & SEO		
Tool	**URL**	**Function**
Google Search Console	search.google.com/search-console	Organic performance, indexing, Core Web Vitals
Google Analytics 4	analytics.google.com	Cross-channel behavior and conversion tracking
Google Business Profile	google.com/business	Local search visibility and review management
Schema.org	schema.org	Structured data vocabulary reference
Screaming Frog	screamingfrog.co.uk	Technical SEO crawl and audit
Ahrefs / Semrush	ahrefs.com / semrush.com	Keyword research, backlinks, competitive intelligence

Paid Advertising		
Tool	**URL**	**Function**
Google Ads	ads.google.com	Search, Display, YouTube, Performance Max
Meta Ads Manager	business.facebook.com	Facebook and Instagram advertising
TikTok Ads Manager	ads.tiktok.com	TikTok advertising campaigns
Google Tag Manager	tagmanager.google.com	Tag and tracking management

CRM & Automation		
Tool	**URL**	**Function**
GoHighLevel	gohighlevel.com	All-in-one CRM, automation, reputation
HubSpot	hubspot.com	CRM, marketing automation, reporting
Salesforce	salesforce.com	Enterprise CRM and pipeline management

Analytics & Reporting		
Tool	**URL**	**Function**
Looker Studio	lookerstudio.google.com	Unified dashboards across data sources
CallRail	callrail.com	Call tracking and attribution
WhatConverts	whatconverts.com	Multi-channel lead tracking and attribution

Creative & Content

Tool	URL	Function
Canva	canva.com	Design and creative production
CapCut	capcut.com	Video editing for social content
Loom	loom.com	Video recording for UGC and founder content

APPENDIX C

90-Day Quick-Start Checklist

DAYS 1–30: FOUNDATION

Infrastructure

- ☐ GA4 installed and verified
- ☐ GTM configured
- ☐ Conversion tracking active (calls, forms, purchases)
- ☐ CRM integrated with ad platforms
- ☐ Google Business Profile complete and optimized
- ☐ Schema markup on key pages

SEO

- ☐ Technical audit completed
- ☐ Intent mapping done for top 20 priority pages
- ☐ NAP consistency verified across all directories
- ☐ Off-page audit completed (citations, links, mentions)

Reputation

- ☐ Review generation system activated
- ☐ Response protocol established (24-hour standard)
- ☐ Duplicate listings identified and removed

Paid

- ☐ Google Ads account structured
- ☐ Meta Pixel and CAPI configured
- ☐ Initial landing pages built and tested
- ☐ First creative batch (6–8 variations) live

DAYS 31–60: OPTIMIZATION

- ☐ Content updated based on Search Console data
- ☐ Off-page authority building underway
- ☐ GEO content structure implemented
- ☐ Negative keyword list expanded
- ☐ Winning creatives identified and scaled
- ☐ Underperforming campaigns paused or restructured
- ☐ Budget incrementally increased on stable campaigns
- ☐ Unified dashboard live
- ☐ Weekly reporting rhythm established
- ☐ CPA and ROAS targets set and tracked

DAYS 61–90: SCALING

- ☐ Authority gap analysis vs. top competitors completed
- ☐ Digital PR outreach producing first mentions
- ☐ GEO content producing AI citations
- ☐ Full-funnel campaign structure live
- ☐ Retargeting campaigns active
- ☐ Creative pipeline producing 3–5 new variations per week
- ☐ 15+ new reviews generated
- ☐ Review recycling into ad creative underway
- ☐ AI summary monitored and accurate
- ☐ 90-day performance review completed
- ☐ Next 90-day growth targets set
- ☐ Scaling decisions made based on data

APPENDIX D

Key Performance Indicators Reference Sheet

KPI	Formula	Target Benchmark
CPA (Cost Per Acquisition)	Total ad spend ÷ Total conversions	Varies by industry and deal value
ROAS (Return on Ad Spend)	Revenue generated ÷ Ad spend	3:1 minimum; 5:1+ target
LTV (Lifetime Value)	AOV × Purchase frequency × Customer lifespan	Business-specific
CTR (Click-Through Rate)	Clicks ÷ Impressions	3–6% Search; 1–2% Display
Conversion Rate (Ads)	Conversions ÷ Clicks	3–8% local services; 1–4% ecom
Quality Score (Google Ads)	Relevance metric (1–10)	7+ target
Organic CTR	Organic clicks ÷ Impressions	3–10% for positions 1–3
Average Position	Mean ranking across keywords	Top 3 for primary keywords
Review Velocity	New reviews per month	4–8/month minimum (local)
Lead Close Rate	Closed deals ÷ Total leads	Industry-dependent
Page Load Speed (LCP)	Time to Largest Contentful Paint	Under 2.5 seconds
Core Web Vitals Pass Rate	% of URLs passing all three CWV	75%+ of URLs passing

APPENDIX E

Structured Data (Schema) Reference

Schema Type	Best For	Key Properties

LocalBusiness	Any local service business	name, address, phone, hours, priceRange, geo
MedicalBusiness	Healthcare providers	medicalSpecialty, availableService, hasMap
LegalService / Attorney	Law firms	areaServed, legalName, serviceArea
Product	Ecommerce	name, image, description, price, availability, review
FAQPage	Content-rich FAQ sections	mainEntity (Question + Answer pairs)
Review / AggregateRating	Review display in SERPs	ratingValue, reviewCount, bestRating
BreadcrumbList	Site navigation clarity	itemListElement, position, name, item
Article / BlogPosting	Editorial and content pages	headline, author, datePublished, publisher
Organization	Brand-level pages	name, url, logo, sameAs (social profiles)

Implementation Note

Implement all structured data in JSON-LD format, placed in the <head> section. Validate with Google's Rich Results Test: search.google.com/test/rich-results. Re-validate after any major website changes or CMS updates.

APPENDIX F

AI Prompt Library for Marketing

Each prompt follows the ROLE-CONTEXT-TASK-FORMAT framework. Customize the [bracketed placeholders] for your business before use.

SEO — Title Tag & Meta Description

Prompt

Act as an expert SEO copywriter. For the page about [topic/service] on a [business type] website targeting [location], write: (1) 5 title tag variations under 60 characters that include "[keyword]" naturally, (2) 3 meta description variations under 155 characters that include the keyword and a clear CTA.

Google Ads — RSA Headlines

Prompt

Act as a Google Ads specialist. The business is [business name], a [business type] in [location]. The target keyword is "[keyword]". Write 15 RSA headlines (max 30 chars each) covering: 5 intent-match, 4 value proposition, 3 CTA, 2 trust, 1 urgency.

Blog Post — SEO-Optimized Outline

Prompt

Act as an SEO content strategist. Create a detailed outline for a blog post targeting "[keyword]" for a [business type] serving [target audience]. Answer search intent completely, include FAQ section, structured for AI extraction.

Email — Nurture Sequence Email

Prompt

Act as an email marketing specialist. Write nurture Email #3 in a 5-email sequence for [business type] targeting [audience]. Previous emails covered [topic 1] and [topic 2]. Provide value around [topic 3], build on prior emails, end with one clear CTA.

Paid Social — UGC Video Brief

Prompt

Act as a paid social creative director. Write a UGC video brief for a [product/service] targeting [audience]. Include: hook script (3 options), core message, proof element, clear CTA, format requirements.

Reputation — Review Response

Prompt

Act as a reputation management specialist for a [business type]. Write: (1) response to a 5-star review mentioning [staff member] (60–80 words, warm), (2) response to a 3-star review mentioning [complaint] (80–100 words, acknowledging concerns).

Analytics — Performance Analysis

Prompt

Act as a digital marketing analyst. Here is last month's data: [paste]. Compare to prior month: [paste]. Identify: 3 most significant positive changes, 3 concerning trends, highest-leverage actions to take next.

GEO — FAQ Content Creation

Prompt

Act as an SEO content specialist focused on Generative Engine Optimization. For a [business type] in [location], create 8 FAQ entries targeting conversational search queries. Each: directly answer in first sentence, 75–150 words, include local specificity.

Campaign Budget Planning Templates

Budget Allocation by Business Stage

Stage	Monthly Range	Primary Allocation	Key Priority
Early / Launch	$1,500–$5,000	60% Google Ads, 30% SEO, 10% Social	Fast demand capture
Growth	$5,000–$15,000	40% Google, 25% SEO, 20% Paid Social, 15% Other	Diversify; scale what works
Scaling	$15,000–$50,000	35% Google, 20% SEO, 25% Paid Social, 10% Video, 10% Other	Full-funnel; testing pipeline
Market Leader	$50,000+	30% Google, 15% SEO, 25% Paid Social, 15% Video, 15% Other	Brand + retention; defend

CPA-Based Budget Calculator

1. Define monthly revenue target → e.g., $50,000

2. Calculate required new customers → e.g., 20 customers at $2,500 avg value

3. Set target CPA (CAC) → e.g., $200 target

4. Calculate minimum ad budget → 20 × $200 = $4,000 minimum

5. Add 20% testing/waste buffer → $4,000 × 1.20 = $4,800 recommended

6. Divide by expected conversion rate → If CVR = 3%, need 667 clicks → $7.19 max CPC at $4,800

The 90-Day Scaling Decision Framework

Performance Signal	Recommended Action	Decision
CPA at/below target for 14+ days	Scale 20%; revisit in 2 weeks	✓ SCALE
CPA within 20% above, stable	Hold; optimize creative & LP	→ HOLD
CPA 20–50% above target	Reduce 15%; audit; eliminate waste	↓ REDUCE
CPA 50%+ above target	Pause; structural issue to fix	■ PAUSE
CPA at target but low volume	Expand keywords, bids, audiences	⊕ EXPAND
Strong ROAS but poor CAC	Check deal mix; high vs low ticket	■ AUDIT

ABOUT THE AUTHOR

John Vincent Kriney

Founder, OptFirst Internet Marketing · Founder, FirstFirms

John Vincent Kriney is a proud father of two sons, **Zane** and **Spencer**, and is a technology and fitness enthusiast.

He is the founder of **OptFirst Inc.**, a Miami–Fort Lauderdale digital marketing firm where he has worked since the early 2000s helping businesses dominate search rankings.

In 2021 he founded **FirstFirms**, a national law firm and med spa marketing agency that brings the OptFirst growth system to two of the most competitive verticals in digital marketing.

He is the author of *The 8 Step SEO Plan* — a practical, step-by-step guide that distills complex SEO strategies into a simple, actionable roadmap for small and medium businesses. *SEO AI Growth System* is the fourth book in his digital marketing series, building on the original *8 Step SEO Plan* (2014), *The Search Engine Marketing Manual* (2017), and *The Online Marketing Manual* (2020).

For years, Kriney has quietly optimized websites behind the scenes — but now he is stepping back into the spotlight, ready to share his tradecraft with aspiring marketers, business owners, and local entrepreneurs from South Florida to Southern California.

Connect with the Author

Web: optfirst.com · firstfirms.com

LinkedIn: linkedin.com/in/johnkriney

Books by John Vincent Kriney: amazon.com — search "John Vincent Kriney"

SEO AI Growth System

2026 Edition · John Vincent Kriney

OptFirst Internet Marketing has managed campaigns across hundreds of businesses — from local service providers to national ecommerce brands and organizations generating over $60 million in year-one revenue.

Every framework in this book comes from real campaigns, real clients, real markets. Tested under competitive conditions. Refined through failure. Validated through results.

John Vincent Kriney is the founder of OptFirst Internet Marketing.

optfirst.com · firstfirms.com

"The system is the advantage."

Made in the USA
Las Vegas, NV
15 November 2026

7aa7399c-2e36-4f6c-83c2-67fb2e31c574R01